WHAT'S IN THE SKY?

CLOUDS

by Thomas K. Adamson

raintree
a Capstone company — publishers for childre

Raintree is an imprint of Capstone Global Library Limited, a company incorporated in England and Wales having its registered office at 264 Banbury Road, Oxford, OX2 7DY – Registered company number: 6695582

www.raintree.co.uk
myorders@raintree.co.uk

Hardback edition © Capstone Global Library Limited 2023
Paperback edition © Capstone Global Library Limited 2024
The moral rights of the proprietor have been asserted.

All rights reserved. No part of this publication may be reproduced in any form or by any means (including photocopying or storing it in any medium by electronic means and whether or not transiently or incidentally to some other use of this publication) without the written permission of the copyright owner, except in accordance with the provisions of the Copyright, Designs and Patents Act 1988 or under the terms of a licence issued by the Copyright Licensing Agency, 5th Floor, Shackleton House, 4 Battle Bridge Lane, London SE1 2HX (www.cla.co.uk). Applications for the copyright owner's written permission should be addressed to the publisher.

Edited by Alison Deering
Designed by Sarah Bennett
Original illustrations © Capstone Global Library Limited 2023
Picture research by Julie De Adder and Svetlana Zhurkin
Production by Katy LaVigne
Originated by Capstone Global Library Ltd

978 1 3982 4796 3 (hardback)
978 1 3982 4800 7 (paperback)

British Library Cataloguing in Publication Data
A full catalogue record for this book is available from the British Library.

Acknowledgements
We would like to thank the following for permission to reproduce photographs: Shutterstock: AlexReut, 6, Binson Calfort, 10, Calin Tatu, 13, Dark Moon Pictures, 8–9, John D Sirlin, 14–15, kikk, 1, KpaTyH, 5, kuruneko, 17, Piyawat Hirunwattanasuk, 7, Studio-M, 18–19, Sunny Forest, 4 (top left) and throughout, Triff, cover, Vasiliy Merkushev, 11, VisanuPhotoshop, 4 (bottom), Wirestock Creators, 16; Svetlana Zhurkin, 20, 21

Every effort has been made to contact copyright holders of material reproduced in this book. Any omissions will be rectified in subsequent printings if notice is given to the publisher.

All the internet addresses (URLs) given in this book were valid at the time of going to press. However, due to the dynamic nature of the internet, some addresses may have changed, or sites may have changed or ceased to exist since publication. While the author and publisher regret any inconvenience this may cause readers, no responsibility for any such changes can be accepted by either the author or the publisher.

Contents

What are clouds? 4

How do clouds form? 6

Where does rain come from? 8

How do clouds move? 10

What colour are clouds? 12

Do clouds make noise? 14

Why are cloudy days cooler than sunny days? 16

What are the different types of clouds? 18

Water condensation 20

Glossary 22

Find out more 23

Index 24

About the author 24

Words in **bold** are in the glossary.

What are clouds?

4

Clouds are masses of water droplets in the air. They also have tiny pieces of dust and ice. The droplets are very small. They are so small that clouds float high above the ground.

5

How do clouds form?

The air around us is full of water we can't see! Water **evaporates** from Earth's surface. It turns into **water vapour** in the air.

Warm air near the ground rises and cools. The water vapour **condenses** onto tiny dust pieces. Those water droplets grow larger. They come together to form a cloud.

Where does rain come from?

The water droplets in a cloud can grow bigger. They might become heavy enough to fall to the ground. That's rain!

Clouds are high enough in the sky that water can freeze into tiny ice pieces. Those pieces can fall as hail, sleet or snow. They are all different types of **precipitation**.

How do clouds move?

The wind pushes clouds along. It can change their shape. Clouds also move water from place to place.

Clouds are a key part of the **water cycle**. Water evaporates from the ocean and the land. It condenses into clouds. The clouds take rain to different places.

water cycle

What colour are clouds?

Clouds are white when sunlight bounces off them. The light scatters in all the colours of the rainbow. Those colours combine to look white.

Some clouds are very tall. Light does not shine all the way through them. These clouds look dark at the bottom.

Do clouds make noise?

Yes! Sometimes very loud noise. Wind moves water droplets inside a rain cloud. They rub against pieces of ice. That rubbing makes an electric charge. That charge releases as lightning.

Thunder is the sound of the lightning. Lightning heats the air around it. In a second, the heated air expands. The air makes loud crackling and rumbling sounds.

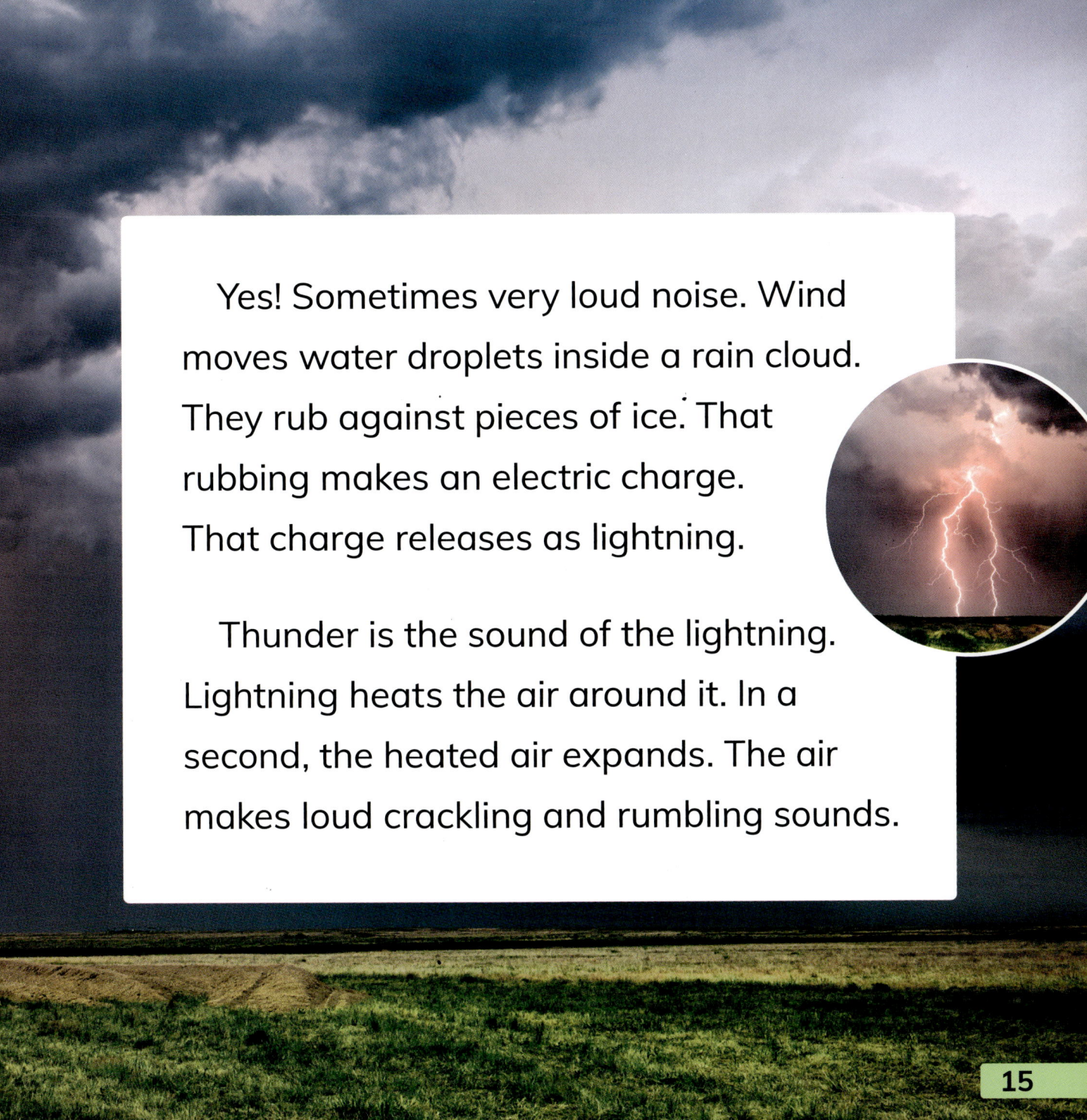

Why are cloudy days cooler than sunny days?

Clouds can block sunlight. That makes the temperature on the ground cooler.

Clouds can also trap the Sun's heat. High, thin clouds trap more heat. Low, thick clouds mostly reflect the Sun's heat.

What are the different types of clouds?

There are many different types of clouds. **Cumulus** clouds look fluffy. **Stratus** clouds are thin and white and cover much of the sky.

cumulus clouds

stratus clouds

Cirrus clouds are thin and wispy. **Altocumulus** clouds are patchy white and grey. **Cumulonimbus** clouds are very tall and cause thunderstorms.

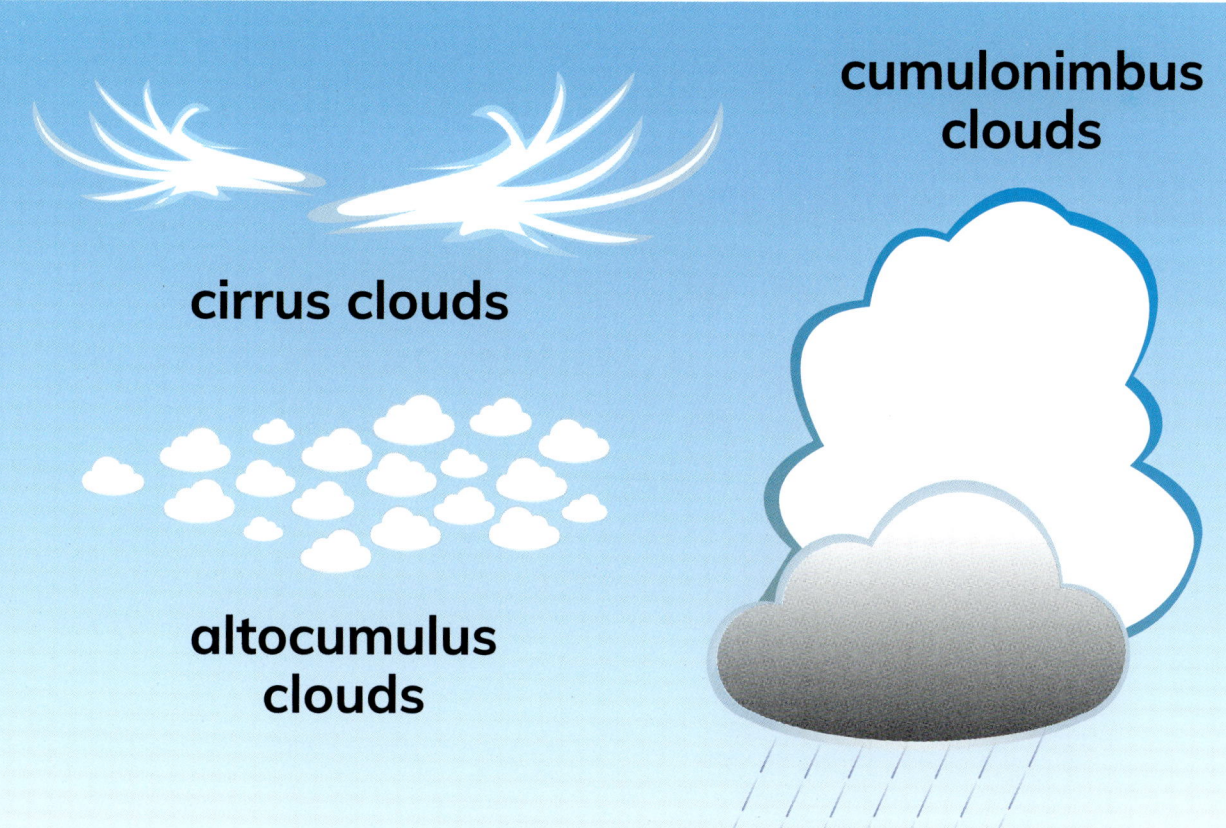

Water condensation

Try this simple activity to see water condensation in action!

What you need

- 2 clear plastic containers with lids
- warm water
- 2 resealable plastic bags
- ice

What to do

1. Fill one plastic container about a third full with warm water. Leave the other container empty.

2. Place lids on both containers.

3. Fill the two resealable bags with ice.

4. Place a bag of ice on top of each container.

5. Watch what happens. After a few minutes, the sides of the container filled with warm water will get foggy as water condenses. The water vapour rises. When it meets the cold air under the lid, the water vapour cools and condenses. You might even see drops of water form on the sides of that container.

6. Remove the ice and open the lid. Can you see how wet the inside of the lid got? Now take the lid off the empty container. How does that one compare?

Glossary

altocumulus high, layered clouds with grey and white ripples

cirrus high, thin clouds made of ice crystals that look like strands of white silk

condense change from gas to liquid; water vapour condenses into liquid water

cumulonimbus huge cloud that can bring hail and tornadoes

cumulus white, puffy cloud with a flat, rounded base

evaporate change from a liquid to a gas

precipitation water that falls from clouds in the form of rain, hail, sleet or snow

stratus low cloud that forms over a large area; stratus clouds often bring light rain

water cycle how water changes as it travels around the world and between the ground and the air

water vapour water in gas form; water vapour is one of many invisible gases in air

Find out more

Books

It's Raining Fish!: Cool Facts About the Weather (Mind-Blowing Science Facts), Kaitlyn Duling (Raintree, 2020)

Totally Amazing Facts About Weather (Mind Benders), Jaclyn Jaycox (Raintree, 2020)

Weather and the Seasons, DK (DK Children, 2019)

Websites

www.bbc.co.uk/bitesize/topics/z6hv9j6/articles/zj3fhcw
Learn more about clouds.

www.dkfindout.com/uk/earth/weather
Find out more about weather.

Index

air 5, 6–7, 15

colours 12

condensation 7

dust 5, 7

evaporation 6, 11

hail 9

ice 5, 9, 15

lightning 15

precipitation 5, 8–9, 11, 15

rain 8, 11, 15

shapes and sizes 10, 12, 17

sleet 9

snow 9

sunlight 12, 17

temperature 17

thunder 15

types of clouds 18–19

water cycle 11

water droplets 5, 7, 8, 12, 15

water vapour 6–7

wind 10, 15

About the author

Thomas K. Adamson has written lots of non-fiction books for kids. Sport, maths, science, cool vehicles – a bit of everything! When not writing, he likes to hike, watch films, eat pizza and, of course, read. Tom lives in South Dakota, USA, with his wife, two sons and a Morkie called Moe.